1ST EDITION

I0409285

HOW TO EVALUATE USED CARS? THE BASIC GUIDE TO A GOOD DEAL!

· VERIFICATION OF LIGHTING, SOUND, AIR CONDITIONING SYSTEMS AND OTHER ELECTRICAL AND ELECTRONIC EQUIPMENT
· EVALUATION OF THE OPERATION OF SAFETY SYSTEMS, SUCH AS AIRBAGS AND SEAT BELTS

CHAPTER 6: VEHICLE HISTORY
· HOW TO OBTAIN VEHICLE HISTORY
· VERIFICATION OF MAINTENANCE RECORDS, ACCIDENTS AND PREVIOUS INCIDENTS
· HOW TO INTERPRET IMPORTANT VEHICLE HISTORY INFORMATION

CHAPTER 7: CONSUMPTION TEST
· HOW TO CHECK THE VEHICLE'S FUEL CONSUMPTION
· EVALUATION OF CONSUMPTION EFFICIENCY IN DIFFERENT DRIVING CONDITIONS
· ENGINE PERFORMANCE VERIFICATION IN RELATION TO CONSUMPTION

CHAPTER 8: CONFIDENCE OF THE BRAND AND MODEL TO BE ASSESSED
· HOW TO EVALUATE THE TRUST AND REPUTATION OF THE VEHICLE BRAND AND MODEL
· ANALYSIS OF THE OPINION OF EXPERTS AND PREVIOUS OWNERS
· VERIFICATION OF AWARDS AND RECOGNITION OF THE VEHICLE ON THE MARKET

CHAPTER 9: VALUE OF THE VEHICLE IN RELATION TO THE MARKET
· HOW TO COMPARE THE PRICE OF THE VEHICLE WITH THE MARKET VALUE AND WITH OTHER SIMILAR VEHICLES
· EVALUATION OF THE VEHICLE'S COST-BENEFIT IN RELATION TO OTHER AVAILABLE OPTIONS
· VERIFICATION OF DISCOUNTS, PROMOTIONS AND OTHER FACTORS THAT MAY AFFECT THE VALUE OF THE VEHICLE

CHAPTER 10: VEHICLE INSURANCE QUOTES
· HOW TO OBTAIN VEHICLE INSURANCE QUOTES FOR THE VEHICLE EVALUATED
· EVALUATION OF PRICES AND COVERAGE OFFERED BY DIFFERENT INSURERS VERIFICATION OF ADDITIONAL

INTRODUCTION

If you are looking for a used vehicle, it is essential that you do a thorough evaluation before making your purchase. After all, a used car can have many unpleasant surprises, such as mechanical or electrical problems, questionable history or inadequate maintenance.

That's why this e-book was created: to help you evaluate a used vehicle efficiently and safely. Through this guide, you will learn to identify possible problems and evaluate important aspects of the car, such as external and internal condition, equipment and vehicle history.

In the first chapter, we'll cover external visual inspection, including tips on how to examine the car's bodywork, windows, tires, and other external parts. Then, in chapter 2, we will detail the internal visual inspection, such as seats, steering wheel, pedals, instruments and other items that must be evaluated before purchase.

In Chapter 3, we'll explain how to evaluate the vehicle's mechanical aspects, such as the engine, transmission, suspension, and brakes. Then, in Chapter 4, we'll discuss the importance of the driving test and how to test it effectively.

In chapter 5, we'll cover electrical, electronic, and security equipment, including checking the lighting system, air conditioning, sound system, and other equipment. In Chapter 6, we'll talk about the importance of vehicle history and how to obtain this information.

In Chapter 7, we'll discuss the importance of fuel testing and how to evaluate a car's performance in terms of fuel efficiency. In chapter 8, we will discuss the reliability of the brand and model being evaluated and how this can affect your purchase decision.

In Chapter 9, you'll learn how to assess the vehicle's market value and how this can influence the fair purchase price. Finally, in Chapter 10, we'll discuss the importance of car insurance quotes, as well as other costs to consider before purchasing a used vehicle.

With this guide, we hope you can feel more confident when evaluating a used vehicle. Always remember that thorough and thorough evaluation can prevent many future problems and ensure that you make a safe and satisfying purchase.

CHAPTER 1
EXTERNAL VISUAL INSPECTION

When looking for a used vehicle, the external visual inspection plays a key role in evaluating the car. In this chapter, we'll explore the key things to consider when looking at your vehicle's overall appearance and identifying potential visible issues, such as dents, scratches, and corrosion.

Overall Vehicle Appearance: When inspecting the overall appearance, it is crucial to assess the quality of the paintwork, the condition of the glass and other external components. In addition to looking for possible damage, such as scratches, dents, or discoloration, it's important to check for signs of previous repairs, such as uneven paintwork or differences in color. These indications may indicate a history of wear and tear or misuse.

Checking for Dents, Scratches and Corrosion: During the inspection of the bodywork, it is recommended to thoroughly check for the presence of dents, scratches or corrosion, especially on the edges and lower parts of the vehicle. These signs could be evidence of a lack of care or proper maintenance. Also, it's important to look out for loose parts, such as bumpers or rear view mirrors, as this could indicate poorly executed repairs or lack of attention to detail.

Assessing Paintwork with a Paint Thickness Gauge: For the most accurate assessment of paint quality, the use of a paint thickness gauge is recommended. This instrument allows checking the thickness of the paint layer applied on the surface of the vehicle. Variations in thickness can indicate paint repairs or partial repainting of the car. It is essential to be aware of these signs in order to make an informed decision about the vehicle's overall condition.

Inspection of Tires, Headlights, Lanterns and Other Visible Items: In addition to the bodywork, it is essential to evaluate the tires, headlights, lamps and other visible components of the vehicle. Check the tires for excessive wear and that they have adequate tread depth to ensure good grip on the road.

Make sure that the headlights and taillights are working properly, and check the validity of the bulbs present in the optical assembly. These items are essential for safety while driving.

Conclusion: The external visual inspection is a fundamental step in the evaluation of a used vehicle. By analyzing the overall appearance, bodywork, paintwork and visible components, you can identify potential issues and assess the vehicle's condition more comprehensively. Remember to use a paint thickness gauge to check the quality of the painting, in addition to checking the validity of fluids, tires and vehicle battery.

CHAPTER 2
INTERNAL VISUAL INSPECTION

The interior visual inspection is a crucial step in evaluating a used vehicle, allowing you to check the quality of materials and interior finish, as well as assessing the overall condition of the car from the inside. In this chapter, we'll improve your ability to evaluate interior materials and finishes, check the condition of seats, carpeting, and upholstery, as well as inspect interior controls.

1. Quality of Materials and Internal Finishing

When evaluating the quality of interior materials and finishes, it is critical to look carefully for signs of wear or damage. Check the seats, dashboard and other surfaces for scratches, stains or discoloration. Also, look for any indications of repairs, such as loose seams or parts that don't fit properly. These signs may indicate excessive wear or even a history of vehicle accidents.

2. Condition of Seats, Carpets and Upholstery

When analyzing the condition of the seats, carpeting and upholstery, focus on identifying possible tears, holes or stains. Check for excessive wear in areas of frequent contact, such as the steering wheel or driver's seat. These areas are more likely to show signs of prolonged use. Also assess the condition of the seat belts, grab handles and other components inside the vehicle.

3. Internal Controls

During the inspection of internal controls, it is important to verify that all equipment is operating correctly. Test the drive buttons, air directing surfaces, interior covers, and all controls such as the steering wheel, gearshift, and instrument panel. Make sure that no buttons are stuck or damaged, and that all functions are operational.

4. Test of All Internal Equipment

It is extremely important to verify that all internal equipment is working properly before closing a deal. Test the air conditioner to ensure it is cooling properly, check the sound system to confirm that all speakers are working and the volume can be adjusted smoothly. Make sure power windows go up and down smoothly, power mirrors adjust properly, headlamps and interior lights are in full working order, and if you have a sunroof, make sure it opens and closes properly. . As you test this equipment, listen for any unusual sounds or abnormal behavior that could indicate problems.

Conclusion

The interior visual inspection is an essential step in assessing the quality of interior materials and finishes, checking the condition of seats, carpeting and upholstery, as well as inspecting a vehicle's interior controls. Make sure all internal equipment is working properly and test them thoroughly before closing. In this way, you will be able to assess the general condition of the vehicle and identify possible problems before making a purchase decision.

CHAPTER 3
MECHANICAL INSPECTION

The mechanical inspection is one of the most crucial steps when evaluating a used vehicle. It is essential to check the condition of all mechanical parts, including the engine, transmission, suspension, brakes and other mechanical systems. In this chapter, we will deepen your understanding of mechanical inspection and the key points to check.

1. Importance of a Specialized Mechanic

To ensure an accurate assessment, it is highly recommended that the mechanical inspection be performed by a skilled and trusted mechanic. An experienced professional will be able to evaluate all components more accurately and identify possible hidden problems. Their expertise is essential for a detailed and reliable analysis.

2. Engine Check

During the engine inspection, it is crucial to check the condition of the bushings, belts, spark plugs and plastic surfaces present in the engine compartment. Assess the cleanliness and wear of these components. In addition, it is essential to examine the combustion chambers for signs of excessive wear, dirt buildup, or other problems. Check for leaks of oil, coolant or other fluids as this could indicate more serious problems.

3. Transmission Evaluation

In the transmission inspection, check for smooth and accurate operation of the transmission, clutch (if manual transmission vehicle) and gearbox. Note if there are strange noises during gear changes or if there is difficulty in shifting gears correctly. These could be signs of transmission issues that need to be investigated.

4. Status of Suspension

When evaluating the suspension, focus on the condition of the shocks, springs and bushings. Check the shock absorbers for oil leaks and that the springs are in good condition with no signs of corrosion or breakage. Bushings should also be examined for excessive wear or looseness. A suspension in good condition is essential for the comfort, stability and safety of the vehicle.

5. Brake Test

Brakes are a crucial part of vehicle safety. It is necessary to carry out a complete test using special equipment to verify the proper functioning of the brake system. This includes assessing pedal pressure, response time and braking efficiency. Also check the status of the brake and cooling fluid, ensuring that they are in proper condition.

6. Other Components and Fluids

When evaluating the vehicle's mechanics, it is important to check the condition of the engine oil, fluids, filters and other relevant parts and components. Check that oil and fluid levels are correct and that there are no signs of contamination or leaks. Also evaluate the condition of the air, oil and fuel filters, as a clogged filter can compromise vehicle performance.

Conclusion

A thorough mechanical inspection is essential to ensure the safety and reliability of the vehicle you intend to purchase. By checking the condition of the engine, transmission, suspension, brakes and other components, you can identify existing and potential problems. Do not hesitate to enlist the help of a specialized mechanic for a more thorough evaluation.

CHAPTER 4
DRIVING TEST

Test driving is one of the most crucial parts of evaluating a used vehicle as it allows you to assess the car's performance, handling and comfort under different driving conditions. In this chapter, we'll deepen your understanding of the driving test and the key things to look out for.

1. Preparing for the Driving Test

Before starting the test drive, it is essential to make sure that the vehicle is in good mechanical and safety condition. Check that all equipment and systems are working properly before starting to drive. This includes the headlights, signal lights, windshield wipers, air conditioning system, sound system and other relevant devices. Also check the tires, their pressure and the presence of any uneven wear.

2. Engine Performance Assessment

During test driving, it is important to evaluate the engine's performance in different rpm ranges and in different gears. Note the smooth, responsive acceleration capability, as well as the ability to pick up speed when passing other vehicles. Pay attention to any engine failures, loss of power, or unusual noises that could indicate mechanical problems.

3. Brake and Stability System Test

During the test drive, pay special attention to the braking system. Test the vehicle's braking ability at both lower and higher speeds to assess the effectiveness and stability of the braking system. Watch for vibrations or steering deviations when braking, as this could indicate problems with the brakes or suspension. Also make sure that the vehicle maintains stability when cornering and that there are no noises or creaks when driving.

4. Comfort and Driving Quality

Evaluate the vehicle's comfort during the test drive. Check the ergonomics of the seats, the sound insulation and the quality of the suspension. Note if there are excessive internal noises, which can indicate wear of finishing parts, suspension, brake or engine. Pay attention to any discomfort caused by excessive vibration or misalignment. Remember that a comfortable vehicle makes for a more enjoyable driving experience.

5. Different Driving Conditions

During the test drive, it is recommended to test the vehicle under different driving conditions. Try to drive on roads, streets with irregular pavement and in more intense traffic conditions. This will help assess the vehicle's ability to handle different situations and provide a broader view of the vehicle's overall performance.

Conclusion

The test drive plays a crucial role in evaluating a used vehicle, allowing you to detect issues that may not be visible during visual or mechanical inspection. By carrying out a comprehensive test drive, you can assess the vehicle's performance, handling and comfort under different conditions. Dedicate enough time to this step and be attentive to all the details to make an accurate assessment of the vehicle before making a purchase decision.

CHAPTER 5
TEST OF ELECTRICAL EQUIPMENT, ELECTRONICS AND SAFETY

When appraising a used vehicle, it is crucial to verify that all electrical, electronic and safety equipment is in working order. In this chapter, we'll explore the essential steps in testing these components.

1. Advanced Security Features Check

Modern vehicles are equipped with a range of advanced safety features like automatic emergency braking system, blind spot monitoring, lane keeping assist and much more. Make sure all these features are present and working correctly. Consult the owner's manual to verify the presence and operation of each feature and test them if possible.

2. Test of Lighting, Sound, Air Conditioning and Other Equipment Systems

In addition to safety features, it is essential to test all of the vehicle's electrical and electronic systems. Check the operation of headlights, flashlights, turn signals, brake lights, reverse lights and fog lights. Test the sound system by checking that all speakers are working properly. Evaluate the air conditioner, verifying that it cools adequately and that the controls are responding correctly. Test power windows, power locks, power mirrors and other similar equipment, ensuring that all buttons and controls are working properly.

3. Assessment of Security Systems

Safety systems such as airbags and seat belts play a crucial role in protecting vehicle occupants in the event of an accident. Check that all airbags are present and in good working order. Test the seat belts by opening and closing them to ensure they are working properly and not slack. Also check that there is no wear on the belts.

We highlight the importance of evaluating safety systems using a scanner capable of identifying error codes or vehicle malfunctions. If you do not have a scanner, we recommend taking the vehicle to a specialized mechanic to carry out this evaluation.

Carrying out a complete test of the vehicle's electrical, electronic and safety equipment is essential to ensure safety and comfort during use. Don't neglect this step when evaluating a used vehicle.

Conclusion

In Chapter 5, we highlighted the importance of testing a used vehicle's electrical, electronic, and safety equipment. Be sure to verify the presence and operation of advanced safety features, test lighting, sound, air conditioning and other electrical equipment, and evaluate the operation of safety systems such as airbags and seat belts. Ensuring all these components are in good condition will contribute to a safe and satisfying driving experience.

CHAPTER 6
IMPORTANCE OF VEHICLE HISTORY IN BUYING A USED CAR

In Chapter 6, we covered the relevance of vehicle history when buying a used car. The vehicle's history contains valuable information that can reveal problems and assist in decision-making during the purchase process. Next, we will expand and enhance the content of this chapter.

1. Obtaining Vehicle History

To obtain the vehicle's history, there are several companies that offer the consultation service online or in person. These companies compile data from different sources such as accident records, maintenance, auctions, ownership history, among others. In addition, it is possible to obtain information on fines and debts directly from the State Traffic Department.

2. Checking Previous Maintenance, Accidents and Incidents Records

When reviewing vehicle history, it is essential to check past maintenance, accident and incident records. These records can reveal crucial information about the vehicle's condition and possible recurring issues. If the car has a long history of repairs and maintenance, this could indicate that it requires more frequent care than you expect.

3. Interpreting Important Vehicle History Information

When analyzing history, it is important to pay attention to certain key information. Check the number of previous owners, as a vehicle with many owner changes can indicate problems. Consider whether the car has been in a major accident or has been recovered from theft or theft, as these events can affect the integrity of the vehicle. Also, check for any outstanding debts or fines, as this could result in legal or financial problems for the new owner.

4. Obtaining a Precautionary Report

We recommend obtaining a precautionary report, carried out by specialized companies, as a complement to the vehicle history. This report provides additional information about the car, including details about the originality of the bodywork, possible tampering, claims history, among other relevant aspects. The precautionary report is an important tool in evaluating a used vehicle and can provide a more comprehensive view of its condition.

Conclusion

In Chapter 6, we discussed the importance of vehicle history when buying a used car. We explored obtaining vehicle history through specialized companies and the Detran, checking maintenance records, accidents and previous incidents, interpreting relevant information and the importance of obtaining an injunction. By considering vehicle history, you are taking steps to avoid problems and ensure a safe and satisfying purchase.

CHAPTER 7
TESTING AND EVALUATING THE CONSUMPTION OF FUEL

In Chapter 7, we covered the importance of considering fuel consumption when buying a used vehicle. Now, let's expand and enhance the content of this chapter, providing more information on how to test and evaluate a car's fuel consumption.

1. Checking the Vehicle's Fuel Consumption

There are several ways to check your vehicle's fuel consumption. One option is to use smartphone apps designed to monitor fuel consumption. These applications record the amount of fuel used during a given period and the distance traveled, thus calculating the average consumption. Another option is to manually record fuel consumption in a spreadsheet, noting the amount of fuel supplied and the mileage traveled.

In addition, there are specialized diagnostic tools such as advanced automotive scanners that can provide detailed information about the vehicle's fuel consumption. These tools allow you to access data from the vehicle's on-board computer, including real-time and historical fuel consumption information.

2. Evaluating Fuel Efficiency in Different Driving Conditions

To obtain a comprehensive assessment of fuel efficiency, it is recommended to test the vehicle under different driving conditions. During the test, it is important to maintain a constant speed in each driving mode, such as on roads, urban streets and highways. Write down the fuel consumption in each situation to compare and evaluate efficiency in different environments.

Remember that factors such as traffic, the topography of the region and the weight of the vehicle can influence fuel consumption. Therefore, it is critical to perform the test under realistic conditions, replicating the type of driving you normally do.

3. Checking Engine Performance in Relation to Consumption

Engine performance also plays an important role in fuel efficiency. During the test, evaluate engine performance against consumption by monitoring acceleration at different speeds. Write down fuel consumption in each situation to identify patterns and compare performance.

It is important to remember that actual fuel consumption may vary according to several factors, such as driving style, type of fuel used, proper vehicle maintenance and even weather conditions. Therefore, it is recommended to perform the fuel test in different driving situations and consider an average to obtain a more accurate assessment of the vehicle's fuel efficiency.

Conclusion

In Chapter 7, we highlighted the importance of testing and evaluating fuel consumption when buying a used vehicle. We explain different methods to check consumption, such as using smartphone apps, spreadsheets and diagnostic tools. We also emphasize the importance of evaluating efficiency under different driving conditions and checking engine performance against consumption. Remember that the test must be performed under realistic conditions for the most accurate results. By considering fuel consumption, you will be making an informed and cost-effective decision when purchasing a used car.

CHAPTER 8
ASSESSING TRUST AND REPUTATION BRAND AND MODEL

In Chapter 8, we covered the importance of assessing the trust and reputation of the make and model of vehicle being considered for purchase. Now, we are going to expand and enhance the content of this chapter, providing more information on how to carry out this assessment effectively.

1. Researching the Make and Model

The first step in assessing vehicle make and model trustworthiness and reputation is to conduct detailed internet research. It is recommended to consult websites specialized in automobiles, forums and communities of owners, where it is possible to find ratings, opinions and comments about the vehicle in question.

When reading these reviews, be aware of aspects such as safety, quality, reliability, performance, comfort and technology. Consider both positive and negative reviews to get a comprehensive view of the vehicle.

2. Expert Opinions

In addition to owner opinions, it is important to analyze the opinion of experts in the field, such as automotive journalists and car experts. These professionals usually perform detailed analyzes of vehicles, highlighting their strengths and weaknesses. Search for reviews and comparisons that address aspects that are relevant to you, such as safety, fuel economy, comfort or drivability.

3. Experiences of Previous Owners

Another valuable source of information is the vehicle's previous owners. If possible, contact them to gain insight into the car's reliability, the quality of maintenance performed, and any issues that may have occurred. These experiences can give you real perspective on the vehicle and help you make a more informed decision.

4. Awards and Recognitions

Checking whether the vehicle has already received an award or recognition from bodies or institutions in the automotive market is also a way of assessing its reliability and quality. Safety, quality, design or energy efficiency awards are positive indicators that the vehicle meets high standards and can be a reliable choice.

5. Additional Considerations

It is important to remember that confidence in the make and model of the vehicle should not be the only factor considered in the purchase decision. It is also essential to assess the general condition of the vehicle, its mechanics and its history through detailed inspections and analyses. This more comprehensive assessment will help ensure that you are making a safe and reliable choice.

Conclusion

In Chapter 8, we highlighted the importance of assessing the trust and reputation of the vehicle make and model. We explain the importance of searching the internet, consulting experts, getting information from previous owners, and checking awards and recognition. However, we emphasize that these assessments must be considered in conjunction with a complete analysis of the vehicle, including its general condition and history, to ensure a reliable and safe purchase.

CHAPTER 9
ASSESSING THE VALUE OF THE VEHICLE IN RELATION TO THE MARKET

In Chapter 9, we discussed the importance of considering the vehicle's market value when evaluating a purchase. We are now going to expand and enhance the content of this chapter, providing more information on how to perform this assessment comprehensively.

1. Comparison with Market Value

In order to compare the price of the vehicle with the market value, it is essential to take into account factors such as the age, mileage and general condition of the vehicle. Websites specialized in buying and selling vehicles, such as the Fipe table, can be useful to provide a price reference.

However, it is important to remember that market value may vary depending on geographic location and supply and demand at the time of purchase. Additionally, some vehicle models may be in greater demand, which may result in higher prices. Therefore, it is recommended to research different sources and consult experts to obtain a more accurate assessment.

2. Comparison with Similar Vehicles

In addition to comparing the market value, it is essential to compare the vehicle with other similar vehicles. This will help determine if the price is reasonable in relation to the vehicle's features and condition. Websites for buying and selling vehicles, as well as consulting opinions from specialists in the field, can be useful in this comparison.

When performing this comparison, consider factors such as age, mileage, overall condition, additional features, and brand and model reputation. This will help to have a clearer view of the vehicle's value in relation to others available on the market.

3. Cost-Benefit Assessment

When assessing the value of the vehicle, it is important to consider the cost-effectiveness in relation to other options available on the market. In addition to the purchase price, consider factors such as fuel economy, performance, reliability, comfort and safety, as well as maintenance and repair costs over time.

A vehicle with a higher starting price but lower fuel consumption and lower maintenance costs may provide better value for money in the long run. It is therefore essential to evaluate all of these aspects to gain a complete picture of the vehicle's value.

4. Discounts, Promotions and Other Factors

In addition to the previously mentioned factors, it is important to be aware of discounts, promotions and other factors that can affect the value of the vehicle. Some dealerships offer discounts on older model vehicles when a new model is introduced, while some automakers offer financial incentives for purchasing electric or hybrid vehicles.

However, it is important to remember that some discounts may be offered at the expense of the vehicle's resale value. Therefore, before making a purchase decision based on discounts and promotions, consider all relevant information and weigh the short-term and long-term benefits.

Conclusion

When assessing the market value of a vehicle, it is crucial to consider the price compared to market value, other similar vehicles and other factors such as value for money, discounts and promotions. Conduct research across different sources, consult experts and consider all relevant aspects to get an accurate assessment. With this information in hand, it will be possible to make a more informed purchase decision and obtain a good cost-benefit ratio.

CHAPTER 10
EVALUATING INSURANCE OPTIONS VEHICLE

In Chapter 10, we discussed the importance of checking vehicle insurance options when evaluating a vehicle for purchase. We are now going to expand and enhance the content of this chapter by providing additional information on how to perform this assessment comprehensively.

1. How to get car insurance quotes for the evaluated vehicle

There are several ways to obtain car insurance quotes for the vehicle being evaluated. In addition to seeking information directly from insurers, through their websites or telephone, it is recommended to use vehicle insurance comparators. These tools let you compare prices and coverage from different insurers in one place, making the process of getting quotes easier.

When requesting quotes, please provide accurate vehicle information such as make, model, year of manufacture, mileage and safety features. In addition, correctly inform your driver profile, including data on age, claim history and vehicle garage location. This information will be used by insurers to calculate the amount of insurance.

2. Evaluation of prices and coverage offered by different insurers

When obtaining car insurance quotes, it is crucial to evaluate the prices and coverage offered by each insurer. Consider the following aspects during the analysis:

· Deductible amount: Check the deductible amount, which is the amount you must pay in case of a claim. Make sure the deductible amount is adequate to your needs and financial possibilities.

· Amount of coverage: Evaluate the amount of coverage offered by the insurance. It is important that the indemnity in the event of an accident is sufficient to cover the damages caused to the vehicle.

· 24-hour assistance: Check that the insurance offers 24-hour assistance, which may include services such as towing, changing tires and mechanical assistance. These services can be of great use in emergency situations.

· Coverage for third parties: It is essential to check whether the insurance offers coverage for damage caused to third parties in the event of an accident. This coverage is important to protect you and your assets against possible legal liabilities.

3. Verification of additional benefits offered by different insurers

In addition to the aspects mentioned above, it is important to check the additional benefits that each insurer offers. Some examples include:

· Reserve car: Some insurers offer the option of a reserve car in the event of an accident. This means you will have access to a replacement vehicle while yours is under repair.

· Coverage for glass: Check if the insurance offers coverage for damage to the vehicle's glass, such as windshield and side windows. This coverage can be useful, as glass is susceptible to damage in everyday situations.

· Coverage for accessories: Check if the insurance offers coverage for accessories installed in the vehicle, such as a sound system, multimedia center, alarm, among others. It is important to ensure that these items are also protected in the event of theft, theft or damage.

Conclusion

When evaluating a vehicle for purchase, it is essential to review the available insurance options and evaluate the prices, coverage and benefits offered by different insurers. Carry out a careful analysis, taking into account the aspects mentioned in this chapter. In this way, you can choose insurance that meets your needs, offers adequate protection for your vehicle and provides peace of mind in the event of accidents or unforeseen circumstances.

CONCLUSION

In this e-book, we provide information and techniques that can help with the purchase of a used vehicle, making this challenging and risky process safer and more informed. Throughout the book, we discuss several fundamental aspects to be considered when evaluating a used vehicle, from reviewing the documentation to carrying out performance and consumption tests.

We emphasize the importance of a complete and thorough assessment, covering vehicle documentation, vehicle history, brand and model reliability, market value and obtaining vehicle insurance quotes. These elements are crucial to assist in decision making and ensure a safer and more satisfying purchase.

It is essential to emphasize that consultation with specialized professionals, such as mechanics and appraisers, is highly recommended to obtain a complete and accurate assessment of the vehicle. These experts have the technical knowledge and experience to identify potential problems and provide valuable guidance during the purchase process.

Buying a used vehicle can be an excellent alternative to saving money, but it is essential to carry out a comprehensive assessment to avoid future problems. By following the guidelines and techniques presented in this e-book, you will be better equipped to make an informed choice that meets your needs and provides you with security and satisfaction as a buyer.

Always remember to research, compare and carry out careful evaluations. Be diligent in checking the vehicle's documents, history, and condition. Also, don't hesitate to seek professional advice when needed. That way, you will be more confident and prepared to take advantage of purchasing a used vehicle.

We hope that the information shared in this e-book is useful and that you make an informed and satisfactory choice when purchasing a used vehicle. Good luck on your journey as a buyer and may you enjoy many safe and exciting adventures behind the wheel!

AUTHOR

Architect and urban planner since 2018, graduated from Centro Universitário Metodista – IPA, in Porto Alegre – RS. Postgraduate degree in Contemporary Education from the Instituto Federal Sul Riograndense in Charqueadas – RS.

Acting as a freelancer in the management and conduction of works and projects, since 2019 as an architect hired by the Municipality of Cachoeirinha - RS, coordinating the real estate cadastre and georeferencing sector. Also conducting works, such as the Pedreira Events Center in Eldorado do Sul, with more than 3000m² of built area implanted in a plot of more than 1 hectare, managing field teams and producing the various projects necessary for the development of the work.

Producer of digital manuals for civil construction, always aiming to provide a practical and easy-to-understand step-by-step, whether for investors or architects/engineers at the beginning of their careers. Seeking to give the reader security in decision-making, clarity in processes and economy of time and resources.

HOW TO EVALUATE USED CARS – THE BASIC GUIDE TO A GOOD DEAL!

stay in touch

Instagram: @rholmerphilipe

Email:
rholmercms@hotmail.com

Portfolio:
behance.net/rholmerphilipe

ISBN: 9798857635629